ISBN 978-1-5277-9206-7
PIBN 10897595

# 1 MONTH OF
# FREE
# READING

## at

## www.ForgottenBooks.com

By purchasing this book you are eligible for one month membership to ForgottenBooks.com, giving you unlimited access to our entire collection of over 1,000,000 titles via our web site and mobile apps.

To claim your free month visit:

www.forgottenbooks.com/free897595

**English**
**Français**
**Deutsche**
**Italiano**
**Español**
**Português**

# www.forgottenbooks.com

**Mythology** Photography **Fiction**
Fishing Christianity **Art** Cooking
Essays Buddhism Freemasonry
Medicine **Biology** Music **Ancient**
**Egypt** Evolution Carpentry Physics
Dance Geology **Mathematics** Fitness
Shakespeare **Folklore** Yoga Marketing
**Confidence** Immortality Biographies
Poetry **Psychology** Witchcraft
Electronics Chemistry History **Law**
Accounting **Philosophy** Anthropology
Alchemy Drama Quantum Mechanics
Atheism Sexual Health **Ancient History**
**Entrepreneurship** Languages Sport
Paleontology Needlework Islam
**Metaphysics** Investment Archaeology
Parenting Statistics Criminology
**Motivational**

# CLEANING AND DEBURRING WORKSTATION
# OPERATIONS MANUAL

by

**Karl Murphy**
**Richard Norcross**
**Peter Tanguy**
**Frederick Proctor**

**June, 1988**

Certain commercial equipment is identified in this paper to adequately describe the systems under development. Such identification does not imply recommendation or endorsement by the National Bureau of Standards, nor does it imply that the equipment is the necessarily the best available for the purpose.

# Table of Contents

# I.   INTRODUCTION

This manual provides instructions for the operation of the Cleaning and Deburring Workstation (CDWS) at the National Bureau of Standards' Automated Manufacturing Research Facility (AMRF). The instruction sets are limited to the normal start-up and shut-down procedures of the workstation enabling an operator to run basic demonstrations and tests.

This section gives a brief description of the workstation's operation, explains how the manual is organized, describes who should use this manual, and enumerates the documentation convention used.

## 1.  WORKSTATION DESCRIPTION

The CDWS cleans and deburrs the metal workpieces or parts that were machined at the AMRF. Equipment at the workstation includes a PUMA 760 robot for deburring, a Unimate 2000 robot for part handling and buffing, a washer/dryer system for cleaning, a buffing wheel system for buffing, a rotary vise for clamping parts, and two roller tray stations for receiving parts. The workstation floor plan is shown in Appendix A.

Research at the workstation has focused on using computerized part geometry to generate accurate robot paths for both part handling and for deburring. At a computer display of a part that requires deburring, an operator selects the edges that the PUMA 760 robot must deburr along with various deburring parameters such as tool speed and feed rate. The operator also specifies how the part is to be gripped by the Unimate 2000 robot, how the part is to be clamped in the vise, and how the part is to be placed in the washer/dryer. These instructions, called process plans, are translated into the proper format and sent to the robots.

The machined parts enter and exit the workstation on part trays placed at the roller tray stations. The Unimate 2000 robot, under command of the workstation controller, grasps a part as instructed by the process plan and carries the part either to the rotary vise or to the washer/dryer system. Normally the process plan specifies deburring, and the Unimate 2000 robot puts the part in the vise where it is clamped. The PUMA 760 robot moves to the vise and begins to deburr the edges on the top side of the part, changing its deburring tool as required. To account for part misplacement and robot inaccuracies, the PUMA 760 robot adjusts the calculated locations of the edges specified in the process plan by detecting the correct location of each point on the path with a wrist force sensor. The Unimate 2000 robot reorients the part and the PUMA 760 robot again locates and deburrs the edges on the newly exposed face. If a part is to be cleaned, the Unimate 2000 places it on the washer/dryer index table. From here the part cycles through the washer, where a hot water spray cleans the part, and through the dryer, where hot air jets dry the part. The buffing wheel system has not been integrated into the workstation operations. When it is, the Unimate 2000 robot will hold parts into the spinning wheels, buffing the edges and faces of the part. When all cleaning and deburring processes are complete, the Unimate 2000 robot transfers the part back to the tray station.

The workstation uses a hierarchical control structure as shown in Figure 1. At the top is the workstation controller (WSC) [1, 2]. The WSC accepts commands from either an operator at the workstation or from a remote source, such as the AMRF cell controller. The WSC sends commands and data to the two robots and to the washer/dryer system. The PUMA 760 robot is controlled by an NBS developed system, the Real-time Control System (RCS) [3, 4]. The PUMA 760 RCS controls the deburring tools and the quick change wrist. The Unimate 2000 robot is controlled by VAL, the vendor supplied controller [5, 6]. The VAL controller sends commands to the tray stations and to the buffing wheels. Both the PUMA 760 RCS and the Unimate 2000 VAL controller can send commands to the rotary vise but only one system has control of the vise at a given time. Arbitration is coordinated by the WSC. The WSC does not control the vise directly because the WSC control delay is too large.

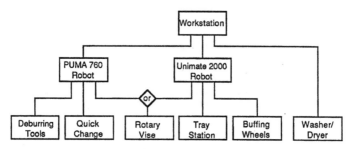

Figure 1. CDWS Control Hierarchy

## 2. HOW THIS MANUAL IS ORGANIZED

This manual describes the power-up and operation procedures needed to demonstrate the CDWS. There are three basic modes of operating the workstation, a stand-alone demonstration, an AMRF integrated demonstration, and a cell and database test.

Chapter II, Demonstration Instructions, is the normal entry point to this manual and provides instructions for the three demonstration modes, stand alone, integrated, and cell/ database test. This chapter leads the operator through the remainder of the manual in the sequence needed for the different demonstrations. The stand-alone demonstration includes the two robots and edge selection but does not require the AMRF cell controller nor the AMRF database to be running. In this mode, the workstation is under the direct control of the operator. See Section II.2. The AMRF integrated demonstration includes both robots and requires that both the AMRF cell controller and the AMRF database are running but does not include edge selections. In this mode, the workstation is under the control of the AMRF cell. See Section II.3. The cell and database test requires that the cell and database are running but does not require the robots. This mode is used when the cell and database operators want to test their systems' interfaces to CDWS (see Section II.4). Safety procedures for running the workstation are provided in Section II.1.

Chapter III, Unimate 2000 Robot, lists in detail the start-up and shut-down procedures for the Unimate 2000 robot. This includes the Unimate 2000 VAL controller and the controllers for the associated equipment.

Chapter IV, PUMA 760 Robot, lists the start-up and shut-down procedure for the PUMA 760 Robot. This chapter also includes general information about the PUMA 760 RCS and certain error handling procedures.

Chapter V, Workstation Controller, lists the start-up and shut-down procedure for the WSC. The commands which can be entered by the operator in the stand-alone mode are also enumerated.

Chapter VI, Part Placement, explains how to manually position parts into the part trays at the roller tray stations.

Chapter VII, Manual Vise Operation, explains how to manually open and close the vise.

Chapter VIII, Process Plan Generation, covers the operator interface to process planning at the workstation. The operator can graphically specify deburring plans by selecting edges on a part along with deburring parameters. Process plans are generated during the stand-alone demonstration mode.

### 3. WHO SHOULD USE THIS MANUAL

This manual is intended for NBS personnel and for government, industry, and university researchers who need to power-up, operate, and shut-down the CDWS or any of its subsystems. It provides step by step instructions that are to be followed at the workstation during power-up, operation, and shut-down. The internal workings and interfaces of the CDWS subsystems are covered in other documents [1, 2; 3, 4, 5, 6]. This manual assumes that the operator only has minimal experience with computers. Terms such as computer terminal, keyboard, and mouse should be understood.

The Cleaning and Deburring Workstation is an ongoing research effort. Several aspects of operation will change without notice. If the operator encounters an error that can not be resolved, he should contact one of the principle operators for assistance. This manual contains more information than is immediately needed by an operator. A suitably trained operator may utilize any aspect of this manual with which he feels comfortable.

### 4. DOCUMENTATION CONVENTION

This manual uses the following conventions:

- The name of each keyboard key mentioned in the text appears in uppercase courier letters. For example, the carriage-return key appears as RETURN .

- In a control-key sequence, a caret (^) represents the CONTROL key. For example, control-C appears as ^C, and is accomplished by holding the CONTROL key down, pressing C, and releasing the CONTROL key.

- Information that appears on the terminal screen appears in bold courier print. For example, the :R rsl> prompt appears on the 760 RCS terminal when the RCS system is awaiting input from the keyboard.

Information that you must enter exactly as it appears in the manual is in plain courier print. For example, "Enter 3 LOAD" means to type 3 LOAD and press RETURN.

Variable data that you must enter is in italics courier print. The instruction "Enter *block #* LOAD" indicates that you are to replace *block #* with a specific block number when you enter the load command. For example, you might type 3 LOAD and press RETURN .

Optional data that you may omit is represented within curly brackets ({ }). Most optional data is also variable data and appears in italics courier print within curly brackets. The instruction "Enter RECEIVE_TRAY { (*Tray_serial_#*) }" indicates that you may either omit the serial # or specify a desired serial #. For example, you might type RECEIVE_TRAY and press RETURN or you might type RECEIVE_TRAY (123) and press RETURN.

When input is to be typed in a specific window or on a specific board, the actual prompt is included in the command string. For example, on the 760 RCS system the instruction "Enter:R Mst> Halt" means to type Halt and press RETURN on the master board.

The name of a button to be pressed or the label of a switch setting is in double quotes. For example: Flip the power switch to "ON".

# II.   DEMONSTRATION INSTRUCTIONS

This chapter provides instructions for operating the workstation and leads the operator throughout the remainder of the manual in the proper sequence. Before running any equipment, the operator should read Section 1, Workstation Safety.

There are three basic modes of operating the workstation, a stand-alone demonstration, an AMRF integrated demonstration, and cell and database test. A stand-alone demonstration includes both robots and graphic edge selection (see Section I.1) but does not require that the AMRF cell controller nor the AMRF database, IMDAS, is running. In this mode, the workstation is under direct control of the operator. See Section II.2. An AMRF integrated demonstration includes both robots and requires that both the AMRF cell controller and the AMRF database, IMDAS, are running but does not include edge selections. In this mode, the workstation is under the control of the AMRF cell. See Section II.3. A cell and database test requires that the cell and database are running but does not require that the robots and equipment are running. This mode is used when the cell and database operators want to test their systems and the interfaces to workstation controller but do not need to actually process parts. See Section II.4.

## 1.   WORKSTATION SAFETY

Before operating workstation equipment, the operator should be fully trained in the workstation's safety procedures. Two of the more important procedures are:

- WEAR SAFETY GOGGLES or safety glasses with side shields at all times while at the workstation or on the shop floor. Safety goggles are available at all entrances to the shop floor. Ensure that all visitors wear the proper eye protection.

- NEVER ENTER A ROBOT'S WORK VOLUME when the arm power is on, signaled by red lights mounted on each robot. The robots' work volumes are marked by red painted floors and are delineated by black and yellow safety tape. Although the robots may appear to move in a predictable manor, a malfunctioning servo valve, for example, could cause the robot to move unexpectedly and at high speed.

EMERGENCY STOP  There are two ways to stop the equipment in the workstation: HOLD and E-STOP. HOLD is a soft stop, used to temporarily stop the motion of one or both of the robots. When the HOLD is reset, the robot(s) continues the task as normal. E-STOP is a hard stop, used in emergency situations to turn off the power to all of the workstation equipment and apply brakes to the robot arms. When E-STOP is reset, the controllers will need to be initialized. An E-STOP is more difficult to recover from but provides a quicker and more reliable method of stopping the robots than HOLD does.

There are several stop boxes (shown in Figure 2) spread about the workstation. Each has a large "E-STOP" button, which stops all workstation equipment; two "HOLD" buttons, one for the PUMA 760 Robot and one for the Unimate 2000 Robot; and three "RESET" buttons, one for the E-STOP, one for the PUMA HOLD and one for the Unimate HOLD. The stop boxes are wired in series so you can press "HOLD" on one stop box and "RESET" on another. There are also "HOLD" and "E-STOP" buttons on the PUMA 760 joystick but they only stop the PUMA 760 and do not stop any other equipment.

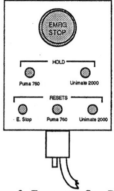

Figure 2. Emergency Stop Box

To 'HOLD' a robot, press the appropriate "HOLD" button on the stop box. It is a good idea to keep your hand near the "HOLD" button when you are issuing new commands. To recover from a "HOLD", press the appropriate "RESET" button, and the robot will continue to move.

To 'E-STOP' the equipment, press the palm-sized red "EMRG STOP" button. This turns off both robots' arm power, the pneumatic deburring tools, the rotary vise, the washer/dryer index table, the water pump, the heater, the buffing wheels, the dust collector, and the roller tray stations.

## 2. STAND-ALONE DEMONSTRATION

1) Clear parts from vise.
   See Section VII.

2) Position part in tray. (tray 1, sector 1 is the default).
   See Section VI.

3) Power Up Unimate 2000 Robot.
   See Section III. 1.

4) Power Up PUMA 760 Robot.
   See Section IV. 2.

5) Power Up the workstation controller (WSC).

See Section V. 1, select "standalone" in step 4 of that section.

6)  At WSC enter "**WSC>** `teach part`"
    where `part` is from step 2 above.
    See Section V. 2.

7)  Select edges and deburring parameters.
    See Section VIII.

8)  The robots will move and deburr your part.

9)  Answer "n" to:
    **WSC>** *j#* **Save Process Plan? <n/number>**
    *j#* n
    **WSC>**

10) Repeat steps 6 through 9 as required.

11) Shut down the workstation.
    See Sections III. 2, IV. 3, and V. 3.

## 3.    AMRF INTEGRATED DEMONSTRATION

1)  Clear parts from vise.
    See Section VII.

2)  Power Up Unimate 2000 Robot
    See Section III. 1.

3)  Power Up PUMA 760 Robot
    See Section IV. 2.

4)  Power Up the workstation controller
    See Section V. 1, select "fully config" in step 4.

5)  Commands for the Workstation are remotely generated.

6)  At demonstration conclusion, shut down the workstation
    See Sections III. 2, IV. 3, and V. 3.

## 4. .   CELL AND DATABASE TESTING

1)  Start-up the workstation controller
    See Section V, 2, select "cell test" or "dbase test" in step 4.

2)  At testing conclusion, shut down the workstation controller
    See Section V. 3.

# III. UNIMATE 2000 ROBOT

This chapter lists in detail the start-up and shut-down procedure for the Unimate 2000 robot. This includes the Unimate 2000 VAL controller and the controllers for the associated equipment.

## 1. START-UP PROCEDURE

1) Between base of Unimate 2000 robot and green/glass partition, lift red-handled circuit breaker to "ON" position to connect power to Unimate 2000 robot controller.

2) At air valves between load/unload tables and base of Unimate 2000 robot, make sure both air valves are turned counter-clockwise to supply air to Unimate 2000 robot gripper and load/unload tables.

3) At air valve next to rotary vise, make sure air valve is turned counter-clockwise to supply air to rotary vise.

4) At Unimate 2000 robot controller TV950 terminal, turn switch on black switch box next to terminal to "2000CRT" mode.

5) At Unimate 2000 Robot Controller Rack:

a) On front of rack, flip power switch to "ON" position to engage fan and CRT.
b) Open front door to rack. The words **UNIM 25B 4-23-82** will appear on the terminal screen and the orange "ATTENTION" light will be on.
c) Pull the red stop button out if necessary and push in button marked "CONSOLE POWER" to connect power to controller console and green button light will come on. Also, the orange/red "HOLD/ERROR" light will come on and the following will be displayed on the terminal screen:

```
* VAL-II BOOT B2K6.2.0 19JUL85 *

LOAD VAL-II FROM FLOPPY (Y/N)?
```

d) Enter N. The following will appear on the screen:

```
VAL-II 2670.2.0 P/N-935F25 H/W-0 S/N-0 20MAR85 23KW
COPYRIGHT 1981 - UNIMATION, INC.
INITIALIZE (Y/N)?
```

e) Enter N. The prompt "." will then be displayed on the screen. The orange "HOLD" light will go off, but the red "ERROR" light will remain on.

6) At RCS Rack for Unimate 2000 Robot:

a) On front of rack, flip power switch to "ON" position to connect power to rack.

b) Open front door to rack. On top side and behind the front panel of Bucket2 interface box, flip switch to "COMP" mode to enable computer interface to equipment.

7) At RCS Terminal for Unimate 2000 Robot:

a) Flip power switch on back of terminal to "ON" position.
b) On reset box on top of terminal, press momentarily the reset button labelled "System Reset" to reset RCS's computer system. The words **pF86-1.4  R45-1.2 up** will appear on the screen.
c) Enter HEX F7 C9 OUTPUT 26F0 30 ERASE 0 CBOOT 0 MBOOT init-cm 2 D>M. If while entering this command string you make a mistake, use the DEL key to backspace and then re-enter the command correctly. After a short time, the system will initialize itself and the workstation status will appear on the screen. The RCS program is now running.

8) At RCS Rack for Unimate 2000 Robot, open front door to rack and look at front panel of Lord F/T Sensor Controller box. If green light marked "SENSOR HEALTH" is off and "FAILURE ALARM" is sounding, press button marked "INITIALIZE" momentarily to reset controller. Make sure the lower toggle switch of the Acme Interface Box has been turned to the "ON" position, then switch the upper toggle switch to the "ON" position to engage the relays.

9) At the Acme Power Panel behind the RCS racks, pull the red and black power lever up to the "ON" position. At the Acme Pushbutton Panel by the buffing wheels, pull the lit red "STOP" button out, and push the green "CONTROL POWER ON" button in. Power will now be supplied to the Acme equipment.

10) At Unimate 2000 Robot Controller Rack:

a) *** Important: Clear Unimate 2000 Robot's work volume of all personnel. The maximum reach of robot is delineated on the floor using safety tape.
b) Turn switch marked "ARM POWER" to "ON" position to provide power to robot and corresponding orange light will come on. After a short time, the red "ERROR" light will go out. Also, the red safety light towards the end of the Unimate 2000 Robot's boom will come on.

11) At Unimate 2000 Robot Controller TV950 terminal:

a) Flip power switch at back of terminal to "ON" position.
b) Turn switch on black switch box next to terminal to "TV950" mode.
c) Enter EX POWER.UP. Place your hand over the red EMERG STOP button in case of error. The gripper should open and the robot should raise slightly, move to its rest position near the trays, and execute its warm-up motions for seven cycles.
d) At the conclusion of the power-up routine, the robot should return to its rest position at the vise, and the message **The 2000 has been reset** should appear on the screen.
e) To run part handling system under control of workstation controller, turn switch on black switch box next to terminal to "SUN" mode. The Unimate 2000 is now running.

## 2. SHUT-DOWN PROCEDURE

1)  At Unimate 2000 Robot Controller TV950 terminal, Enter DO MOVE REST.VISE. Place your hand over the red EMERG STOP button in case of possible malfunction or operator error. If robot was resting at the load/unload tables, the robot will swing to assigned rest point. If robot was resting at the vise, the robot will stay at assigned rest point. Robot is now at power down position.

2)  At Unimate 2000 Robot Controller Rack, turn switch marked "ARM POWER" to "OFF" position to disable power to robot and corresponding orange light will go out. Also, the red safety light towards the end of the Unimate 2000 Robot's boom will go out. Push the red stop button to turn off the controller, shut the rack, and turn the rack switch to the "OFF" position to disable the fan and CRT.

3)  At RCS Terminal for Unimate 2000 Robot:

    a) Press the space bar to abort the status program. Enter HALT.
    b) Enter BYE to shutdown the RCS's storage disk. Blinking stars should fill the screen, indicating the system has been halted.
    c) On back of terminal, flip power switch to "OFF" position.

4)  At RCS Rack for Unimate 2000 Robot, turn the top toggle switch of the Acme Interface Box to the "OFF" position, then turn the power switch on front of rack to "OFF" position to disconnect power to rack.

5)  At Base of Unimate 2000 Robot, push red-handled circuit breaker to "OFF" position to disconnect power to Unimate 2000 Robot controller.

6)  At Acme Pushbutton Panel, push the red "STOP" button in to cut power to the equipment. At the Acme Power Panel, throw the power lever down to the "OFF" position.

CDWS Operations Manual

12

# IV. PUMA 760 ROBOT

This chapter lists the start-up and shut-down procedure for the PUMA 760 Robot and includes general information about the PUMA 760 RCS and certain error handling procedures.

## 1. START-UP PROCEDURE

\*\*\* Warning: **Do not initialize** the PUMA 760 at power-up. That is, if you see the question: **INITIALIZE (Y/N)?** on the terminal in the PUMA 706 robot controller rack, enter N. If you enter Y and initialize the controller, you will erase all programs and locations stored in VAL. If this happens, see Section 4, Error Handling.

1) Turn on the PUMA 760 Robot Controller Rack; move the large red switch in back to "ON".

2) If the controller is off, hold "AUTO START" button in while you turn on the controller chassis. Continue holding the "AUTO START" button until you hear a click, about 3 sec. Continue with step 3.

   If the controller is already on, a) press the "ARM POWER OFF" button and then b) hold the "AUTO START" button in while you toggle the "LSI INIT" switch (on the top of the chassis). Continue holding the "AUTO START" button until you hear a click, about 3 sec.

3) Clear people from the robot work volume.

4) Press "ARM POWER" button.

5) Press the "JOINT" button on the PUMA joystick and move the robot to the vise-safe location. (The tool is vertical, the tool tip is midway between the quick change rack and the robot base, and the tool tip is slightly higher than the top of the quick change rack. See Figure 3.)

\*\*\* Warning: The robot joints will move about 10 degrees during the following step. Keep a hand near the "ARM POWER OFF" or "E-STOP" button.

6) Press the "COMP" button on the PUMA joystick. The robot will move as it calibrates itself.

7) Set the terminal switchbox to "RCS", if not already there.

Figure 3. Vise-Safe Location

8) Turn on: the 760 RCS rack, all the chassis in the rack, and the RCS terminal.

9) Press the black "SYS Reset" button on the terminal switch box. If not already set, switch the black rotary switch to "MST", and set the toggle switch to "TERM".

*** Warning: The robot will move to the vise-safe location during the following step. Make sure it is free to move, or keep your hand on the HOLD SET button.

10) Press function keys f9 then f10. Or, if the terminal was off, enter:
    HEX F7 C9 OUTPUT 26F0 30 ERASE 0 CBOOT 0 MBOOT 4 D>M
If you make a typing mistake while entering this command, use the DEL key to backspace and re-enter the command correctly.

After the disk spins up, the computer will begin to print messages from the various boards. The robot will move to the vise safe location. The tip of the robot tool should be higher than the top of the quick change rack. If not see errors below. You should end up on **Mst>** with the message: **\*\*\* System Running \*\*\* DONE.**

The 760 RCS system is now running and ready to receive commands from the SUN.

## 2. SHUT-DOWN PROCEDURE

1) Move arm to a safe location and press the ARM POWER OFF button.

2) Enter **Mst>** BYE. This stops the RCS controller, unloads the disk, and starts "snake" (the screen saver program).

3) Turn off all equipment that you turned on.

## 3. GENERAL INFORMATION

MOVING FROM BOARD TO BOARD: The RCS system has 7 levels running on a separate computer boards. There is a master board, **Mst>** , and 6 slave boards, **rcs>, task>, path>, prim>, joint>** and **comm>**. Although there are 7 boards, there is only one terminal which "talks" to all 7 boards.

To move from board to board use the terminal function keys. The RCS terminal has 11 function keys labeled f1 thru f11. Each has a separate function as labled on the duct tape. For example, f3 will move you to the **Mst>** board. You can move back and forth between any two boards.

To operate the RCS system you must enter different things on different boards. In this manual, the prompt tells you the board to enter data on. For example: **Mst>** GO. means type a G, an O, and RETURN on the **Mst>** board. On the computer screen, the prompt shows the name of the board you are on. Some commands entered on the master board will automatically send commands to slave boards, ex: **Mst>** GO.

JOYSTICK: Enter either **Mst>** JOYSTICK or **task>** 912 LOAD. These commands clear certain errors and enable the joystick. To move the robot, flip the enable switch up and set the velocity to a low value. (Rotate clockwise.) The most common motions are tool X, tool Y and tool Z. See Figure 4.

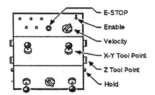

Figure 4. PUMA 760 Robot Joystick

STATUS: Enter **Mst>** STAT, **task>** STAT, or **path>** STAT. The status of the robot will be displayed. The task and path status are different and the master status is the combination of both task and path status.

COMMAND DIRECTORIES: **task>** 910 LIST and **task>** 920 LIST. These blocks list different commands to the robot. For example, entering: **task>** 924 LOAD will command the robot to move to VISE-SAFE.

SUN COMMANDS: **task>** ENABLE-SUN-CMD and **task>** DISABLE-SUN-CMD tells TASK level to accept/ignore commands from the workstation controller.

VISE: **path>** GET-VISE locks the 2000 out of the vise volume. **path>** DROP-VISE drops the vise volume request. The 2000 can now get the vise. The vise volume arbitration is handled by the RCS for the Unimate 2000.

RSL: At start-up, the RSL board is told to accept robot paths from the SUN. This program is called: **rsl>** BGO. If RSL is to ignore the paths from the SUN enter: **Mst>** ABT_RSL **rsl>** DGO. You will not get a prompt back. To abort these processes, BGO or DGO, enter: **Mst>** ABT RSL.

## 4. ERROR HANDLING

Most errors can be corrected by doing step (a) below. If the joystick is needed, see Section 3, General Information.

a) General error recovery:
- Enter **Mst>** GO. Then, if there is no error:
- Enable joystick and move the robot near the vise safe location.
- Enter: **t&p>** 924 LOAD. The robot will move to the vise safe location. The tip of the robot tool should be higher than the top of the quick-change rack. If not see error (d) below.

b) Force sensor is down.
- **comm>** INIT-FORCE.
- Check power to sensor chassis, cable connections, etc.

c) The robot does not move
- It may already be at vise-safe. Try **t&p>** 922 LOAD. The robot should move to qc-gate.
- Press the HOLD CLEAR button on the RCS joystick.
- See General errors

d) The robot does not move to vise-safe location, the tool is lower than the top of the quick change. This generally means that the robot does not know that it has a tool.
- **t&p>** 911 LOAD, commands reset, then **t&p>** 924 LOAD.
- Check power to bucket-1 interface chassis, cable connections, etc.

e) Quick Change Error:
  If the robot has an error over the quick change and stops moving:
  **t&p>** X-UNLOCK **Unlock quick change?** Y
  **t&p>** 911 LOAD. This restarts RCS. **t&p>** 912 LOAD. This activates the joystick

f) No air: tools don't work, quick change doesn't work
- Turn on air valve feeding the PUMA 760 robot.

g) No Auto Start program on VAL.
- **Do not initialize!**
- Do steps 1-5 in Section 1, START-UP PROCEDURE above <u>without</u> pressing the "AUTO START" button. The robot will not move.
- Flip the two toggle switches on the terminal switch box to "LOCAL" and "TERM".
- Enter CAL. Answer Y. The robot will move.
- Continue on step 7 in Section 1, START-UP PROCEDURE above.

# V. WORKSTATION CONTROLLER

This chapter lists the start-up and shut-down procedures for the WSC. The commands which can be entered by the operator in the stand-alone mode are also enumerated. All commands in this chapter are enter by using the SUN terminal, keyboard, and mouse.

## 1. START-UP PROCEDURE

1) If SUN terminal is running "Game Of Life"; press any key and go to step #4.

2) Otherwise, turn the SUN computer on via the three main power switches located on the SUN disk pedestal, the SUN base pedestal, and the back of the SUN terminal screen.

3) Log into the system with: **cdws login:** rjn

4) Start the controller program by:
    open rootmenu,
        position the mouse arrow over the gray background area.
        hold down the right mouse button.
    open "WSC" submenu,
        with right button depressed, move mouse to the "=>" to the right of "WSC"
    highlight appropriate configuration with mouse arrow.

| | |
|---|---|
| w/ prompts | controller prompts for CMM & robot status |
| fully config | CMM at "demo", database & both robots running |
| CELL test | CMM at "demo", database & robots not running |
| dBase test | CMM at "demo", database running, robots not running |
| standalone | no CMM, both robots running |
| 760 only | no CMM, Unimate 2000 not running |
| cmm | (ignore) |

5) Wait. The system takes about three minutes to initiate and run.
    The prompt WSC> will appear in the terminal window.

## 2. TERMINAL INPUTS

To enter data into the terminal window the mouse arrow must be in that window.

If the prompt, **WSC>,** is missing the system may still accept inputs. However a prompt can be generated by entering: ()

Commands are entered by typing the command followed by its parameters. E.g.:
    **WSC>** teach block_fh 1 1 0

When the system asks a question it will provide an identifier which must be the first element of

your response. Also, the system generally gives a guide to the acceptable responses. For example:
```
WSC> j5 Save this Process Plan? <n/number>
WSC> j5 n
```

## 3.  SHUT-DOWN PROCEDURE

1)   Enter: **WSC>** quit

2)   Wait. "Game of Life" will appear on the screen.

3)   If desired, turn off the three power switches.

## 4.  COMMON COMMANDS

The following is a list of common commands to the WSC. These are listed as a reference and are not needed by a novice user. Commands to the workstation can come from the cell or can be typed into the workstation control window. The first three commands are the standard sequence from the cell.

**RECEIVE_TRAY   {TRAY_SER_NR} {TRANSFER_POINT}**
   locks the transfer point tray station and fetches the *Tray Contents Report* from the database IMDAS.
   TRAY_SER_NR is the "tray serial number" and defaults to "mt_tray".
   TRANSFER_POINT is the AMRF name for the tray station and defaults to "CDWS_TP1".

**DEBURR_LOT     LOT_ID {OPERATION_SHEET} {LOT_TYPE}**
   deburrs the lot of parts as specified by the instruction_set declared in the operation_sheet.
   LOT_ID is the lot's identification number and is a required entry.
   OPERATION_SHEET defaults to the standard instruction_set numbers.
   LOT_TYPE is the part description and defaults to "MIXED" which means "do all parts".

**SHIP_TRAY      {TRANSFER_POINT}**
   releases the transfer point tray station and clears those parts from the internal database.

**teach part-type {sector} {tray} {location}**
   develops and tests a process plan for the specified type of part.
   part-type is the part description and is a required entry.
   sector is the tray sector where the part is placed and defaults to "1".
   tray is the tray station where the part is placed and defaults to "1".
   location is the current "vise" position of the practice part and defaults to "TRAY".

**move-part     part goal**
   moves the named part to the goal position. The part must be in the internal database.
   part is the name of the part and is required.
   goal is either "TRAY" or a vise-grip location where the part is to be placed and is required.

18

**add-part**     *part part-type tray-station {location}*
     adds a part to the internal database.
     part is the name of the part and is required.
     part-type is the part description and is a required entry.
     tray-station is the buffer position of the part (e.g. TRAY11,TRAY24) and is required.
     location is the current position of the part and defaults to the buffer position.

**set-location** *part location*
     changes the part's location in the internal database (w/o part movement).
     part is the name of the part and is a required entry.
     location is the part's new position.

**RCS**      *command-string*
     sends the command to the task level of the RCS controller controlling the PUMA 760.
     command-string is the command to be sent and must be enclosed in double quotes (" ").

**VAL**      *command*
     sends the command to the VAL+ controller controlling the Unimate 2000.
     command can be multiple words and doesn't require quotes (e.g. VAL do depart 150).

**re-initialize_VAL**
     re-establishes the command/status handshake with the Unimate 2000. This command should be executed when the 2000 is returned to the WSC's control after being in local mode.

**deburr-part**    *part {instruction_set}*
     deburrs the specified part based on the instruction_set.
     part is the name of the part and is a required entry
     instruction_set is the process plan to be used and defaults based on the part type

**deburr-loops** *part-type loops*
     commands the PUMA 760 to deburr the specified loops. A workpiece of the given type must be properly positioned in the vise and the appropriate RSL file downloaded to the PUMA's RCS controller. E.g. deburr-loops pipeclamp_fv (0 1 2)
     part-type is the workpiece description and is a required entry
     loops is the list of loop numbers which correlate to the RSL file

**RCS_status**
     returns the most recent status from the RCS controller controlling the PUMA 760.

**initialize_dbase**
     generates the UVA protocol for establishing communications with the AMRF database.

**set db-flag nil**
     shifts the controller to look at local copies of data rather than the AMRF database, IMDAS.

19

**`initialize_CELL`**
establishes communications with the CMM and in turn with the remote command source.

**`set CMM nil`**
causes the controller to ignore the remote command source.

**`initialize_RCS`**
establishes communications with the RCS controlling the PUMA 760.

**`initialize_VAL`**
establishes communications with the VAL+ controller controlling the Unimate 2000.

**`set VAL-flag nil`**
breaks the communications with the VAL+ controller controlling the Unimate 2000.

# VI.  PART PLACEMENT

The Cleaning and Deburring Workstation is designed to receive parts at two tray transfer stations. Parts arrive on predefined trays with the parts in preset locations. The workstation does not have the capability to acquire randomly placed parts. Therefore, to operate the workstation in a stand-alone mode, an operator must position parts accurately by using the part positioning guides.

The first step in positioning parts is to identify the tray and sector number. Figure 5 shows the numbering convention used at the CDWS. Sectors "1" through "4" reference the corners of the trays. Since sector "5" has no such reference, it should be avoided when possible.

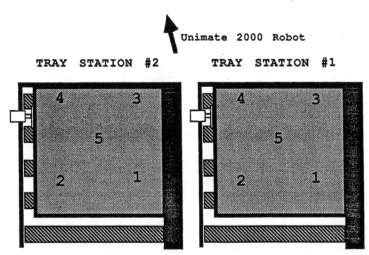

Figure 5.  Tray Numbering Convention

To assist proper part positioning in a sector, positioning guides are provided for the operator. Each part has a unique positioning guide, as shown in Figure 6. The guides are kept under the tray stations. To use the guides:

1) Turn off the Unimate 2000 arm power. See Section III. 2.

2) Orient the guide in accordance with the instructions on the guide.

3) Place the guide in the reference corner for the desired sector.

4) Place the part in the hole in the guide.

21

5)  Carefully lift the guide over the part.

The robot can capture the part when it is less then roughly one-half inch off the true position. Therefore, the guide may nudge the part when being removed.  Also, if the guides are missing, a educated guess will probably suffice.

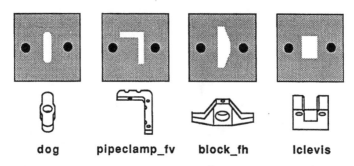

| dog | pipeclamp_fv | block_fh | lclevis |

Figure 6.  Part Positioning Guides

The basic demonstration at the CDWS involves the teaching and deburring of a single part.  The command for this demonstration contains default parameters for tray "1", sector "1".  Figure 7 shows the proper location of the default position.

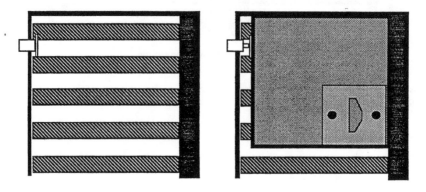

Figure 7.  Default Part Position For WSC "teach" Command

22

# VII. MANUAL VISE OPERATION

During initialization and during some error recoveries the part fixturing vise requires manual operation. The location of the vise controls is the "Bucket 2 Interface" box. This control box is in the "RCS RACK for UNIMATE 2000". See Workstation Floor Plan. Figure 8 shows the vise control logic. The vise must be returned to computer control before continuing a demonstration (i.e. the "comp/man" switch, see Figure 8, must be in the "comp" position).

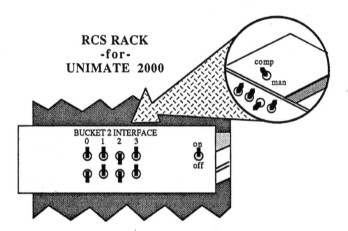

| CONDITION | SWITCH #0 | SWITCH #1 | SWITCH #2 |
|-----------|-----------|-----------|-----------|
| VISE @ 0 | DOWN | DOWN | DON'T CARE |
| VISE @ 90 | DON'T CARE | UP | DON'T CARE |
| VISE @ 180 | UP | DOWN | DON'T CARE |
| VISE OPEN | DON'T CARE | DON'T CARE | DOWN |
| VISE CLOSE | DON'T CARE | DON'T CARE | UP |

Figure 8. Manual Vise Controls

To clear a part from the vise:

1) Turn off the Unimate 2000 arm power. See Section III. 2.

2) Turn off the PUMA 760 arm power. See Section IV. 3.

3) Make sure air is supplied to the rotary vise by turning the air valve next to the vise counter-clockwise.

4) Turn on the RCS Rack for Unimate 2000 Robot.

5) Put "comp/man" switch to "man" position. See Figure 8.

6) Put switch #2 in the down position to open the vise.

7) Remove part from the vise.

8) Return "comp/man" switch to the "comp" position.

# VIII. PROCESS PLAN GENERATION

This chapter covers the operator interface to process planning at the workstation. The operator can graphically specify deburring plans by selecting edges on a part along with deburring parameters.

During the execution of the "teach" command (See Chapter V.) the workstation's process plan generator is automatically initiated. The user interface to the plan generator is a set of graphic window displays. The master window, shown in Figure 9, will appear on the screen. Using the mouse, open the command menu (by pressing the far right mouse button) and select "select deburring". The other menu selections such as "add vise grip" and "add robot grip" are beyond the scope of this document. See [2] for explanation.

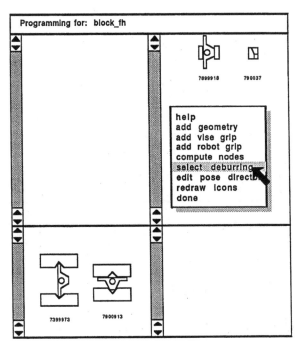

Figure 9. Process Planning Window.

When the edge selection window appears, shown in Figure 10, select the edges to be deburred (by pressing the left mouse button) along with the desired tools, tool parameters, and views. In this window, the mouse locator is the "printer's fist" which initially appears in the middle of the window display.

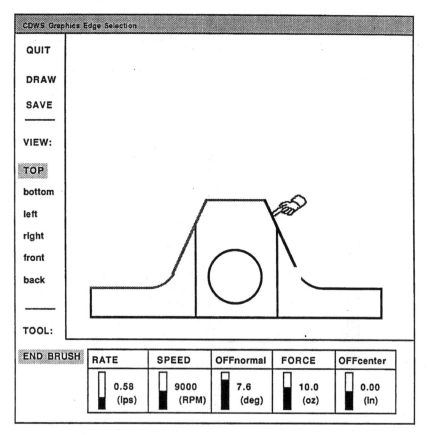

Figure 10. Process Plan Generator's Graphic Display Window

The workstation controller continues to function during process plan generation. To perform other tasks the user moves the mouse back to the terminal interface window and enters the command. The graphic window will not automatically redraw itself. If the graphic display is obscured by another window, bring the graphic display back to the top by pressing the function key L5 and

26

select one of the views with the mouse.

### View Selection:
To select a view, place the mouse pointer (the "printer's fist") over the name of the desired views, and click the left mouse button. Only views with "word" names can be fixtured by the workstation. Other views (e.g. x=1) can be displayed on the screen but are meaningless to the robots.

### Tool Selection:
To select a different tool, point the mouse at the current tool and click the left mouse button. The system toggles through the available tools in a round robin. The current suite of tools is; a 120 grit aluminum oxide end brush, a 320 grit aluminum oxide hole brush with 5/8 inch outside diameter, and a carbide chamfering tool which is listed as a spindle.

### Parameter Selection:
Each tool has unique parameters and default settings. When a new tool is selected the parameters and their current values are shown at the bottom of the display. Parameter values are changed via the thermometer graphs at the left of each value. The values change with a single click of the left mouse button.

### Edge Selection:
Edges are selected with the same point and click procedure described above. An edge may be unselected in the same manner in which it was selected. Figure 10 shows sample edge selections.

### The Save Command:
In the upper left corner of the display are three commands, the bottom command saves the users selections in a temporary file. Currently there is no way to correct selections once they are saved, so caution is advised. A "SAVE" selection only saves the edge selections visible at that time, though there may be more then one "SAVE" on each side. The tool and parameters are saved with the edges selected.

### The Draw Command:
The "DRAW" command is only used when generating local geometry files and should be ignored.

### The Quit Command:
The "QUIT" command terminates the session with the plan generator. Upon termination the Process Plan Generator writes an AMRF-formatted process plan based on the saved edge selections. When completed, the display is automatically erased and the workstation controller is advised of the completion. The "teach" command then resumes its normal execution.

27

# APPENDIX A
# WORKSTATION FLOOR PLAN

Cleaning & Deburring Workstation
Boundary

Material Handling Workstation

Kardex System

Robot Work Volume

Scale - feet

0  1  2  3  4

# LIST OF REFERENCES

[1]  Norcross, R.,"CDWS Workstation Control Structure", IEEE International Conference on Robotics and Automation, Philadelphia, PA, 1988.

[2]  Norcross, R., "CDWS Workstation Controller Reference Manual", National Bureau of Standards, to be published.

[3]  "The NBS Real-time Control System / User Reference Manual". Editors: Kilmer, R. and Leake, S.  To be published 1988.

[4]  Murphy, K., "RCS on the CDWS PUMA 760", National Bureau of Standards, to be published.

[5]  Programming Manual, User's Guide to VAL-II, Version 2.0, 398AG1, Unimation, December 1986.

[6]  Proctor, F., "CAD Directed Automatic Part Handling", National Bureau of Standards, to be published.

# READER COMMENT FORM

## CLEANING AND DEBURRING WORKSTATION - OPERATIONS MANUAL

This document is one in a series of publications which document research done at the National Bureau of Standards' Automated Manufacturing Research Facility from 1981 through March, 1987.

You may use this form to comment on the technical content or organization of this document or to contribute suggested editorial changes.

_____

_____

_____

_____

_____

_____

_____

If you wish a reply, give your name, company, and complete mailing address:

_____

_____

_____

_____

What is your occupation?  _____

Note: This form may not be used to order additional copies of this document or other documents in the series. Copies of AMRF documents are available from NTIS.

Please mail your comments to:    AMRF Program Manager
National Bureau of Standards
Building 220, Room B-111
Gaithersburg, MD 20899

ERFORMING ORGANIZATION (If joint or other than NBS, see instructions)

**ATIONAL BUREAU OF STANDARDS**
**EPARTMENT OF COMMERCE**
**ASHINGTON, D.C. 20234**

BSTRACT (A 200-word or less factual summary of most significant information. If document includes a significant ibliography or literature survey, mention it here)

his manual provides instruction for the operation of the Cleaning and Deburring orkstation at the National Bureau of Standards' Automated Manufacturing Research Facility. The instruction sets are limited to the normal start-up and shut-down procedures of the workstation enabling an operator to run basic demonstrations and tests.